Managing fluvial and coastal environments

GW00602774

Rosemary Charlton
Julian Orford

6 5 4 3 2

© Rosemary Charlton and Julian Orford
   2002

Designed by Colourpoint Books.
Printed by Nicholson & Bass Ltd.

Cover: The location is the mouth of the River Shimna at Newcastle, Co Down. The slide was taken from a small plane in 1983; the view is approximately looking west-north-west. It shows the building of hard protection (rock armouring) along the Newcastle seafront, which also fixed the mouth position and hence the flood discharge capacity of the Shimna. The potential of the Shimna to back up at time of flood and high tide is reflected in the small artificial lake that occupies the small flood plain, the latter proving to be too hazardous for the encroachment of Newcastle. The fixing of the Shimna mouth controls the peak amount of river flood discharge and may contribute to the flooding in the lower Shimna at times of extreme discharge. (© J Orford)

Rosemary Charlton is a Lecturer in the Department of Geography, National University of Ireland, Maynooth, where she teaches undergraduate and postgraduate courses in hydrology, fluvial geomorphology and water resource management. She is currently carrying out research to predict future water resource availability in Ireland under a changing climate.

Julian Orford is Professor of Physical Geography in the School of Geography at Queen's University, Belfast. He started lecturing at Queen's in 1977, after doing postgraduate research on coastal sedimentation at Salford (MSc) and Reading (PhD). He lectures on coastal environments in terms of their physical development and human management. His research now concentrates on coastal morphological response to both sea-level change, and storminess related to climatic change. He has been Head of the School of Geography, QUB, since 2001.

ISBN 1 898392 87 0

Colourpoint Books
Unit D5, Ards Business Centre
Jubilee Road
NEWTOWNARDS
County Down
Northern Ireland
BT23 4YH
Tel: 028 9182 0505
Fax: 028 9182 1900
E-mail: info@colourpoint.co.uk
Web-site: www.colourpoint.co.uk

Chapter 1

Human interaction in coastal and fluvial environments

Uses and demands upon river and valley zones

Throughout the history of civilisation, water has been of fundamental importance. It was not until humankind first started to manipulate and redistribute water in time and space that the earliest western civilisations appeared over 5000 years ago on the floodplains of the Nile in Egypt and the Tigris and Euphrates in Mesopotamia (present day Iraq). They are described by archaeologists as 'hydraulic civilisations' because of their almost total dependence on river water. The ability of the first engineers to divert river water to fields to cultivate crops meant that a move away from subsistence was possible for the first time.

Through time our knowledge and understanding of rivers and freshwater resources has developed, often through trial and error. Over the last century mechanisation has allowed water to be manipulated on a far greater scale than ever before. Huge dams regulate the flow of most of the world's larger rivers, vast expanses of cropland green the desert and the flow of water supplying the largest cities is exceeded only by a few major rivers. At the same time, river and floodplain environments are becoming increasingly built-up as urban areas expand, more land is cultivated and industrial development takes place.

Water supply and increasing demand

Domestic and municipal

The global demand for water is increasing. Population growth accounts for some of this increase but the effect is greatly magnified by other factors. Urbanisation – the drift towards cities that is especially marked in LEDCs – is a major factor because urban areas have different patterns of water consumption and a higher per-capita demand. When considering water use in urban areas, domestic use is often linked with **municipal use** – water used by shops, businesses, schools, hospitals, public buildings and parks. Whereas the global average water consumption in rural areas is 100–200 litres per person per day, the average for urban areas is 300–600 litres. To meet this demand many cities depend on **inter–basin transfers**, where large volumes of water are imported to the area from outside the local drainage basin. Water may be transferred over hundreds of kilometres.

Industrial

Industrial use accounts for a quarter of global water withdrawals. This water is used in many ways: for electricity generation, cooling, transportation, washing, as a solvent and may also form part of the finished product. The process of industrialisation is a major influence, with increasing amounts of water required as industrialisation progresses. In North America, water withdrawals for industrial use increased 20 times during the twentieth century.

Agricultural

Two-thirds of global freshwater withdrawals are used in agriculture, mainly for irrigating cropland. Today approximately 16% of the world's cropland is irrigated, with almost half the global harvest coming from this irrigated area. Some of the most fertile soils are to be found in arid and semi-arid environments and irrigation makes it possible to bring these new areas into production. The greatest expansion, known as the 'Green Revolution', occurred between 1950 and the late 1970s. During this period the cultivated area more than doubled (from 94 million to 230 million ha). Without irrigation water drawn from rivers and aquifers Egypt would be unable to produce food and yields from the critical grain producing regions of the US western Great Plains, northern China and north-west India would be significantly reduced.

Waste and water pollution

Increasing demand for water leads to a corresponding increase in the volume of wastewater. Water-borne disease, a primary health concern in LEDCs, is spread by pathogens (disease-causing organisms) in untreated domestic waste. Industrial effluent may contain heavy metals and other substances that are toxic in very small concentrations; irrigation drainage water contains high concentrations of salts and agro-chemicals. When a certain volume of contaminated water is discharged into a water body, many times that volume can become polluted. This reduces the available water supply and has obvious implications for downstream users and freshwater ecosystems. In both MEDCs and LEDCs shortages of fresh water and the increasing pollution of water bodies are becoming limiting factors in economic and social development.

Non-consumptive use of rivers, lakes, reservoirs and valley zones

As well as providing a supply of water, rivers, lakes and reservoirs are used in other ways: commercial shipping, fisheries and recreation. Large rivers such as the Rhine, Danube and Mississippi are important for navigation and extensive channel engineering works have been carried out to modify channels for shipping. Recreation is of increasing importance and takes many forms, for example fishing, sailing, windsurfing, canoeing and walking. Rivers, lakes and reservoirs also have scenic value, providing opportunities for tourism development.

Floodplain development

Floodplains have always been prime sites for agricultural, urban and industrial development as they provide a supply of water, a source of power, fertile soils and flat land for agriculture, development and communications. Some of the highest population densities in the world (greater than 100 people per km^2) are on the irrigated floodplains of the Indus, Ganges, Brahmaputra, Huang-ho and Yangtze. In both MEDCs and LEDCs irrigation techniques are often inefficient, leading to salinisation of soils and rivers and the abandonment of land. In wetter climates, the drainage of wetland areas for agriculture has reduced habitat diversity and brought about management problems.

The growth of urban areas also puts increasing pressure on floodplain environments and has a number of direct and indirect impacts. The hydrology of the area is affected as less water infiltrates the paved area and drains and sewers rapidly transport floodwater. At the same time less water is able to infiltrate and percolate downwards to replenish groundwater. Flood alleviation works to protect urban and some agricultural areas from flooding can have major impacts on river and valley zones.

The increasing complexity of human interactions in coastal environments

Society, technology and coastal occupancy

Human use of coastal environments is dependent both on society's cultural perspectives on the value and use of the coastal zone, and on the technological and economic basis by which society can successfully occupy one of the most hazardous of the Earth's natural environments. Human occupation of a zone that is exposed to extreme conditions (storms and floods, coastal erosion and infrastructure loss) requires overriding reasons for remaining there, as well as a sophisticated level of technology to support its occupation. Society needs something from the coastal zone that cannot be provided elsewhere at the same cost, while the probability of something hazardous happening can be accepted given the need for these required resources, although original location reasons can be lost over time and people remain through inertia. Coastal dwellers trust technology will provide protection from threats from the restless sea and their occupation is only possible where society has also the financial capability to protect coastal property.

Contrasting society responses to similar coastal hazards

Comparing the damage from intense tropical storm activity in the USA (hurricanes) and Bangladesh (cyclones) identifies the differing response of each society to a similar physical process (storm surge), although against different coastal morphology. In the USA modern technology allows advanced and effective warning of hurricanes followed up by horizontal evacuation of people from low-lying areas and/or vertical evacuation in strengthened hurricane-resistant buildings. Property losses can be covered by insurance, so life is more important than property. In Bangladesh property and land occupancy cannot be forfeited even under pressure from deadly cyclones. There is no insurance support and there are few safe places to run to, so evacuation is rarely a viable option. Land claim is essential to subsistence cropping and despite cyclone flooding threatening to family existence, they must stay huddled where they live and work. Clinging to trees is the major lifesaver at times when thousands will drown. Societies respond to the same physical challenges differently through their varied cultural and economic structures.

Pre-industrial society and the coast

Human occupancy of the coastal zone (at least in the western world) has progressed in parallel with the technological development of society. For example in England, prior to its nineteenth century urban-industrial expansion, the coast was merely the limit of the terrestrial world and its land-based activities. Coastal use was tied to the exploitation of fish and for shipping, and humans continually adjusted to the demands of the sea. Coastal sites often suffered severe damage from storms: Dunwich in Suffolk is now a small coastal village, yet in the thirteenth century rivalled London as England's major port, a position it lost due to continual coastal erosion. In the past, society's impact on the shoreline was small though potentially important as it formed the core for much growth during the last two centuries.

Coastal activity associated with ports and harbours began with ancient civilisations (eg Egyptian, Greek and Roman) when coastal trading was the means of supporting and holding Mediterranean empires together. This theme reappeared in the empire and colonial phases characteristic of post-sixteenth century Europe and was dependent on trans-oceanic connections. The vibrancy of ocean connections ensured that many LEDCs have major coastal urban areas based on old colonial trading ports (eg Rio de Janeiro, Cape Town, Bombay, and Manila); some MEDCs have similar colonial coastal connections (eg New York and Sydney).

There were other coastal problems that constrained settlement even away from the threat of coastal storm and flood. Freshwater and saltwater coastal marshes, a common feature of upper reaches of estuaries and tidal inlets, were often associated with debilitating diseases due to poor water quality when human waste (pre sewerage-controls) contaminated watercourses. Shakespeare wrote of the 'ague', a form of malaria that was common around the Thames and Essex marshes in medieval times. Coastal lowlands could only be opened up to safe agricultural development with drainage and protection from sea inundation. The Dutch were the most successful in developing forms of protection for coastal wetlands from the fourteenth century: words like **groyne**, **dyke** and **bund** all derive from the Dutch and their slow and painstaking reclamation of intertidal land (**polderisation**) by which salt was washed out to convert such areas into productive agricultural land. The English Fens and The Wash shoreline were substantially altered by Dutch-designed reclamation from the sixteenth century. Europe saw further expansion of estuary reclamation for agriculture in the nineteenth century; Ireland used this approach as a basis for famine relief, with Youghal estuary (Co Cork) and Wexford Harbour substantially reclaimed. These projects are small in comparison to the Dutch reclamation of the Zuider Zee in the twentieth century.

Industrial society and the coast

The rise of industry and associated urbanism in Victorian Britain indirectly generated spectacular changes in coastal zone occupancy that became the model for the industrial world. As a consequence of conscious improvement in workers' conditions by some industrialists, holiday time (later leisure time) became available by the 1870s. Holidays were for health purposes and coastal bathing and breathing coastal air was thought to be beneficial, so it followed that holidays were to be spent by the seaside.

Victorian railways supplied cheap and quick transport to the coast and the late nineteenth century saw an explosion of seaside resorts, exemplified in the rumbustious gaiety of Blackpool for the working class of northern Britain and dignified Scarborough and Bournemouth for the more image- and moral-conscious middle classes. From small beginnings, coastal resorts experienced some of the highest population growth rates in Britain between 1851 and 1911.

The attraction of British resorts is easy to explain: a week of freedom from the Victorian straitjacket of moral and social inhibition was a heady lure to workers living the rest of the year in grinding poverty. The rapid expansion of coastal towns with its emphasis on the shoreline meant that static infrastructures of buildings, jostling for the shoreline, were located virtually overnight on changing coastlines, planting the seeds of coastal management problems.

By the 1960s relative disposable wealth meant that people were now able to retire to places traditionally associated with fun and freedom. Numerous estates of retirement bungalows, crowding the coastline for a perspective on the sea, have sprung up. Rising income, cheap air travel and the pursuit of the 'S5' factor (sun, sand, sea, surf and sex) has. since the 1970s. undercut traditional UK resorts and promoted those in warmer climes (eg Spain derives 10% of earnings from coastal tourism). The decline of UK coastal resorts mirrors their equally rapid rise, and underlines a changing contemporary coastal ideal that does not now include the sedate pleasures ('S2' — sand and sandwiches) central to UK resort existence.

Post-industrial society and the coast

New images of ocean living have become accepted as the core of modern leisure and the basis of a near universally desired life-style. The shoreline in the twenty-first century is the new beacon drawing the masses back from the heartlands of continents and their decaying industries and problematic cities. There is a globalisation of this coastal message ('Go to the coast!') that overrides local views of coastal living. It is estimated that 37% of the world's population now lives within 100 km of the coast with expansion rates greater than any other area. However, the ability of the coastal zone to sustain this expansion is in doubt.

Chapter 2

Modern approaches to river management

The need for channelisation

Channelisation is the term used to describe modifications made to the width, depth and plan form of natural stream channels. Modifications are also made to control channel stability where excessive erosion or deposition is occurring at a specific location or along a channel reach. Numerous rivers worldwide have been affected by channelisation. **Resource 1** illustrates some of the techniques used.

Flood control and land drainage

The periodic inundation of floodplains by rivers is a natural process but certain land use change and the impacts of channel engineering can increase this tendency. Channel modifications are frequently carried out to increase flow velocity and channel capacity for

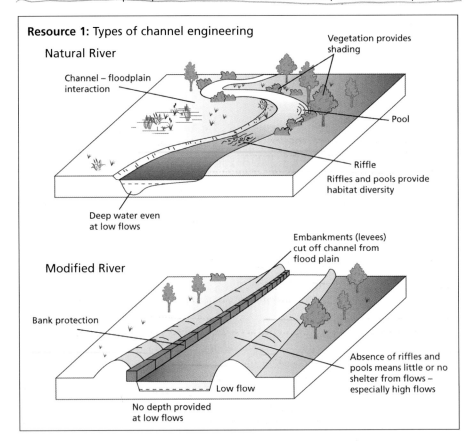

Resource 1: Types of channel engineering

Natural River

Vegetation provides shading

Channel – floodplain interaction

Pool

Riffle

Riffles and pools provide habitat diversity

Deep water even at low flows

Modified River

Embankments (levees) cut off channel from flood plain

Bank protection

Absence of riffles and pools means little or no shelter from flows – especially high flows

Low flow

No depth provided at low flows

flood control. The cross-sectional area can be enlarged by **resectioning**, where the channel is widened (by excavation) and deepened (by dredging) as shown in **Resource 1**. **Realignment** or straightening is usually carried out at the same time. The removal of meanders increases channel gradient and therefore flow velocity. In order to increase flood protection for urban and some agricultural areas, **flood embankments** or **artificial levees** may be constructed at the edge of the channel to contain high flows.

Channelisation for land drainage also involves resectioning and realignment of the main channel and tributaries. Following this, field drains are installed to transport water to the river channels. Land drainage can increase the flood risk further downstream. So over the last 50 years most of the rivers and a number of tributaries in Northern Ireland have been channelised for the dual purposes of land drainage and flood alleviation.

Navigation
For larger rivers where shipping is important, realignment is often carried out to reduce travel distances. Channel deepening increases the navigable length of river and other structures such as weirs and locking systems to control flow may be installed.

Bank protection
The channelisation works described above often cause localised instability as the river attempts to readjust to its natural form. Therefore it is usually necessary to protect the bed and banks using **revetments**. These erosion control structures may be made from a variety of materials, the most common being concrete, sheet piling, riprap (loose boulders or concrete blocks) or gabions (metal baskets filled with rocks).

River training
This is a technique where structures are installed to 'train' the river to erode in some places and deposit in others and thus to develop and maintain a new channel form. River training was used on the Mississippi in order to deepen the channel for navigation. **Resource 2** illustrates this diagrammatically. Spur dykes were built at right angles to one bank to trap sediment. This concentrated erosion at the bed, and the opposite bank was protected from erosion by lining it with mattresses of giant concrete blocks joined by steel cables. This produced a navigable route for shipping and increased flow velocity for flood protection.

The impacts of channelisation

Problems of instability and maintenance
Channelisation often leads to instability along the engineered section as well as upstream and downstream from it. Regular maintenance and/or further channel modifications are necessary in order to keep the channel in its unnatural state. Annual maintenance on the Mississippi, the most engineered channel in the world, costs $180 million a year due to the sheer scale of the engineering works. The resectioned channel is bordered by thousands of kilometres of flood embankments and was shortened by 240 km for flood control and shipping, between 1928 and 1942, using explosives to blast through the necks of meanders. This created major instability problems, with increased erosion occurring upstream from the engineered section as the river

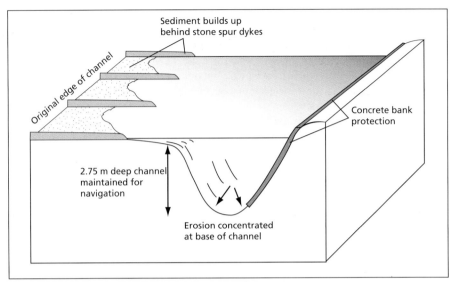

Sediment builds up behind stone spur dykes

Original edge of channel

Concrete bank protection

2.75 m deep channel maintained for navigation

Erosion concentrated at base of channel

Resource 2: River training on the Mississippi

attempted to reduce its gradient. As a result, an increased volume of sediment entered the engineered reach, causing extensive deposition in the form of bars. The straightened river also began to meander again, necessitating further bank protection.

Increased flood hazard

It has often been argued that the problem of flooding has been made worse as a result of channelisation. Speeding up the passage of floodwater and confining flow between embankments to protect one location can cause problems further downstream. When flow is concentrated in an engineered channel there is no floodplain storage, which means that the flood peak is greater and the lag time is greatly reduced. It has been suggested that the extensive channel engineering works and thousands of kilometres of flood embankments may have worsened the floods that occurred on the Mississippi in 1993.

Ecological impacts and aesthetic considerations

In natural channels a range of flow conditions are observed, such as the variations in depth, sediment size and water velocity associated with riffles, pools and meander bends. This natural variation provides a range of habitats for the different types of flora and fauna that live, feed and breed within the river, at the base of the channel or in the riparian zone bordering the channel. A number of species, including most fish, depend on several different environments to complete their life cycle.

- Channel engineering greatly reduces the range of habitats found in rivers and riparian zones.
- Continued maintenance, such as dredging, prevents habitats from re-establishing themselves.

- The removal of undercut banks and boulders means that engineered channels provide little shelter and few resting-places for fish and invertebrates and at high flows the velocity can be too great for some species to withstand.
- At low flows increased temperatures are observed in widened (and therefore shallower) channels, which often lack any kind of shading due to the removal of trees and bank-side vegetation.
- The installation of embankments cuts rivers off from their floodplains, with the consequence that important hydrological, sedimentological and ecological interactions are greatly reduced.
- From an aesthetic viewpoint, heavily engineered channels are rarely attractive in appearance. It is natural features like meander bends, sparkling riffles, deep pools and wetland environments that give a natural river its scenic and amenity value.

The engineering solution versus the need for habitat protection

There is a growing need to develop rivers and water resources but, at the same time, increasing concern about the sustainability of traditional approaches to river management. Over the last two decades new developments have been made in environmentally sensitive and sustainable river management. This applies an understanding of the physical processes controlling channel shape and dimensions while considering all aspects of the river and its floodplain.

Environmentally sensitive approaches to channel engineering

There are a several environmental approaches to land drainage and flood management. These have been successfully carried out over a range of different scales and are illustrated diagrammatically in **Resource 3**.

- Where it is necessary to increase channel capacity, **partial dredging** can be used where channel deepening is limited to a central section. This retains a range of habitats and allows the flora and fauna remaining on the relatively undisturbed parts of the bed each side to re-colonise the deepened section.
- Increasing attention has been given to river margins. These form important habitats, are important in controlling subsurface flows, contribute organic matter to rivers, provide storage for floodwaters and enhance the aesthetic and amenity value of the river. Where space allows, the construction of **distant flood embankments** that are set back from the river channel, allows the river to inundate its floodplain while still providing flood protection for the surrounding area. This causes minimum disruption to the river-floodplain environment and has the added advantage that distant flood embankments do not need to be as high as adjacent embankments to provide the same level of protection.
- Where space is more limited, a **multi-stage channel** can be created. Part of the floodplain adjacent to the river is excavated to increase channel capacity at high flows while leaving the river channel itself unaltered.
- Instability can be controlled using **soft engineering**. Depending on the degree of protection required, different materials can be used, including grass or reeds, geo-textiles or woven fences using willow or wood. Many of these structures encourage

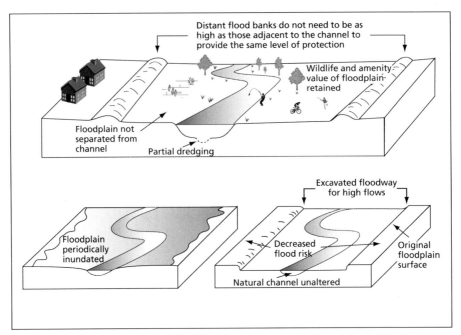

Resource 3: Environmentally sensitive channel engineering techniques

the deposition of sediment and colonisation by vegetation, which reduces their appearance and creates new habitats.

River restoration

This refers to the reinstatement of natural features to an engineered channel in order to improve ecological diversity and aesthetic value. Using structures to encourage scour and deposition at intervals along the channel creates riffles and pools or, for larger channels, pools are excavated mechanically. New meandering channels are excavated with bank protection using soft engineering. Careful calculations are required during the planning stage to ensure that the correct dimensions are selected for meander wavelength and riffle-pool spacing, according to channel dimensions and flow conditions. Although the restoration possibilities for heavily engineered channels are often limited the impact of some structures can be lessened using fast growing species of reeds and willow to create new habitats.

Case study: The Ballysally Blagh, Coleraine

The Ballysally Blagh project was carried out in 1994–5 and was the first flood protection scheme in Northern Ireland where river restoration was one of the main aims. The Ballysally Blagh is a tributary of the Lower River Bann and drains agricultural land before flowing through an urban area. The channel is between two and four metres wide and in places the river has been heavily engineered using rock armouring.

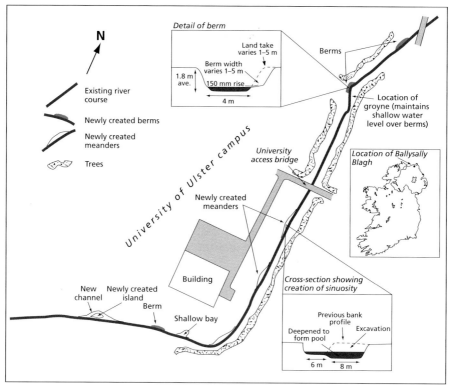

Resource 4: Map of Ballysally Blagh case study

A map of the project reach is shown in **Resource 4**. The river flows through a narrow corridor bordered by residential areas (upper reach) before continuing through the grounds of the University of Ulster (lower reach). In the 1980s property damage occurred as a result of several flood events that affected houses and developments constructed since the 1950s. The main objectives of the project were to:

- provide flood alleviation for the surrounding area;
- improve the visual appearance by reinstating channel features lost during previous drainage works;
- enhance the wildlife, fishery and conservation value;
- develop the river restoration skills of the in-house workforce;
- provide an example for other environmentally sympathetic channel engineering projects.

Upper Reach: Developed Corridor

Flood protection works were carried out along the upper reach by lowering the bed by 500–750 mm, widening the channel by between one and one and a half metres and installing gabions for bank stabilisation. At the same time a number of measures were taken to reduce the impact of the engineering works. These included low-level wet berms, groynes and stone blocks to provide variation in depth, shelter at high flows, and to encourage localised deposition. Flow deflectors were installed to encourage scour and deepening at certain points. Trees and shrubs were also planted along the banks and bird nesting tunnels installed. The appearance of the gabion baskets was improved by planting them with willow stakes.

Resource 5: Preworks

The widening of the channel and introduction of boulders and groynes has caused silt deposition to occur at certain points and a narrower sinuous low-flow channel has formed. This has improved the appearance of the monotonous straight and

Resource 6: Remeander work looking upstream

trench-like channel created by the original engineering works carried out several decades ago, and has increased habitat diversity.

Lower Reach: University Grounds

There was more space available for modification of the lower reach (**Resource 5**), which allowed a number of channel enhancements to be made. Two large 'meanders' were excavated (**Resource 6**) and an island and bay created. The banks were re-profiled and trees planted. Small ledges, low fish weirs and gravel riffles were also installed to increase habitat diversity and enhance channel appearance.

Within a year the island was becoming colonised by vegetation and silt was beginning to accumulate. The island is now a stable feature. The wavelength of the installed meanders was too long in relation to the channel width and as a result narrower, more sinuous meanders have developed within the engineered channel. Despite this, the project has been very successful overall and has achieved the original objectives as well as being a valuable learning experience in the application of these relatively new techniques.

Chapter 3

Basin Management

Aims

At many different scales there are large variations in the distribution of water, land and people over both space and time. Basin management aims to integrate the many conflicting demands made on rivers and water resources, to provide protection from floods and to maintain water quality. The main aims of drainage basin management are listed below:

- the supply, treatment and distribution of water to industrial, domestic and municipal users;
- the supply of water for irrigation;
- hydro-electric generation;
- flood protection;
- water quality control;
- fish and wildlife conservation;
- management and maintenance of channels for navigation;
- to provide recreational facilities / manage recreational activities that might impact on water quality.

Conflict of interest

There is often conflict of interest between different users, for example in California there are continuing disputes between farmers and city planners over the allocation of water. Conflicts also occur between different regions. Although the drainage basin provides the most logical way of dividing the land surface into zones for water resource management, very different areas are delimited by administrative, regional and political boundaries. In fact 47% of the Earth's land area (excluding Antarctica) is covered by **international drainage basins** that are shared between two or more nations. Numerous treaties and water agreements exist to apportion water between different regions within shared basins, for example the Rhine. However, in many parts of the world international agreements are often inadequate or entirely lacking. There is, for example, a long history of controversy over water allocation along the River Nile, which is shared between nine nations.

Planning for future demand

Long time-scales are involved in the design, construction and operation of a project and planning for future trends in supply and demand is of great importance. This can be difficult where information is lacking and major problems have resulted from a lack of hydrological information, uncertain socio-economic contexts and poor environmental assessments. Future changes in climate are predicted to radically alter the distribution and availability of useable water supplies, with serious implications for new and existing basin management schemes.

Traditional strategies

Dams have been constructed for thousands of years to control floods and supply water for irrigation and domestic use. The first large multi-purpose dams were constructed in the US in the early 1930s and, together with water treatment plants, pipelines, aqueducts and sewage systems, enabled the integrated management of rivers and water resources. The success of the large dams at the centre of these projects paved the way for numerous structural developments worldwide, especially between the early 1950s and the 1970s. The scale of development has increased considerably with numerous large-scale projects in LEDCs. There is growing public concern about the negative impacts of large dams and plans for the Ilisu Dam on the Tigris in south-eastern Turkey were shelved in 2001 after a sustained campaign by human rights and environmental groups. However, controversial projects, such as the Three Gorges Dam on the Yangtze in China, are still going ahead in the ongoing race for economic development. Today two-thirds of the world's rivers are regulated to some extent. The beneficial outcomes and impacts of traditional basin management are now considered.

Beneficial Outcomes
- *Flood protection*: Floodwater can be stored in reservoirs for later release, protecting downstream areas and allowing floodplain development.
- *Flow regulation*: Water supply is ensured to meet demands during dry periods. Peak flows, for example snowmelt runoff, which would previously have been 'lost' down the river, can now be stored in reservoirs.
- *Industrial development*: Hydroelectric generation provides a source of energy which, together with a readily available water supply, allows industrial development to take place.
- *Cheap source of electricity*: Once constructed, hydroelectric schemes provide a very cheap source of energy, which is renewable and non-polluting. Revenue from the sale of hydroelectricity can eventually offset construction costs.
- *Irrigated agriculture*: The supply of water for irrigation enables an expansion in the cultivated area and increased agricultural productivity.
- *Water quality control*: Water quality can be managed by regulating flow to ensure that there is a sufficient volume of water to dilute downstream discharges of effluent.
- *Pollution incidents*: In the event of a major pollution incident, water can be released from a reservoir to dilute and flush the polluted water downstream.
- *Recreation*: Reservoirs provide opportunities for recreation and tourism development.
- *Navigation*: Flow regulation can be beneficial for navigation.

Impacts
- *Population displacement*: Major reservoirs inundate a considerable area of land and often involve the displacement of large numbers of people. In India millions have been displaced by the reservoirs created behind 3,300 major dams over the last 50 years. It is estimated that the controversial Narmada project, involving the construction of numerous dams along the 1,200 km Narmada River, will displace over 200,000 people. The Sardar Sarovar dam alone has displaced an estimated 10,000 people with most being inadequately compensated, if at all.

- *Health implications*: The large surface area of standing water in reservoirs increases the risk of water borne disease in tropical countries as it provides an ideal breeding ground for disease-carrying organisms.
- *Cost of construction*: Large engineering projects require vast amounts of capital investment and loans are often taken out by poorer nations to finance construction. However, once projects are completed many do not produce the predicted returns, adding to the economic burden of loan repayments.
- *Environmental problems associated with irrigation*: Water-thirsty crops such as rice and cotton are often cultivated in dryland environments for which they are not suited, using inefficient irrigation techniques. It is estimated that 75–90% of this water is lost due to high evaporation rates. This has resulted in the salinisation of soils with thousands of hectares of fertile soil lost every year. Drainage water contains high concentrations of salts and toxic agro-chemicals.
- *Ecological losses*: Flow regulation changes the nature of ecosystems found in river and wetland environments, migrating fish are unable to pass dams and habitats are destroyed by reservoirs. Irrigated agriculture also has serious adverse effects on ecosystems.
- *Increasing political conflict*: The large-scale manipulation of water that is now possible has the potential to exacerbate existing political tension in water-scarce regions.

New approaches to basin management

Over the last thirty years philosophies of basin management have started to change, as the environmental, social and economic impacts of large structural schemes have become increasingly apparent. Growing public concern for wildlife and landscape protection has also contributed to these changes. **Sustainable basin management** aims to consider the drainage basin as a whole, taking into account water, land use, development and conservation. For a scheme to be sustainable it should provide for the needs of the present without causing deterioration to the river environment, economy or availability of water in the future. Improvements in scientific knowledge mean that it is increasingly possible to work in harmony with natural processes rather than against them. Some examples of sustainable management are outlined below.

- *Land-use management*. Land-use change within the drainage basin can affect runoff and sediment transfer. Ideally, land use and water management should be integrated, although this may not be possible in social, economic and political contexts.
- *Environmentally sensitive approaches to channel engineering*. The techniques discussed in Chapter 2 make it possible to work with the river and are increasingly used instead of hard engineering.
- *Small-scale projects*. Instead of constructing large dams on major rivers, smaller dams can be built on tributaries to provide water and power for local communities.
- *Floodplain zoning*. It is sometimes possible to control the encroachment of urban areas onto floodplains by zoning at-risk areas and restricting development within these zones. This reduces the flood risk and therefore the need for flood engineering.
- *Conservation of water in urban areas*. Water conservation can reduce the need for, and scale of, further structural developments. In New York during the mid 1990s water

meters, low-flush toilets and low volume showerheads were installed in every household by the water company. The consequent reduction in demand meant that the construction of a new reservoir could be postponed for several years.

- *Water-efficient irrigation*: Huge evaporative losses are associated with traditional irrigation techniques that flood fields or furrows with water. Sprinkler and drip irrigation techniques are considerably more efficient and greatly reduce problems of waterlogging and salinisation.

Case study: The Colorado River

Location and environment
The Colorado rises in the Rocky Mountains and flows 2,550 km through parts of seven western states and Mexico to reach the sea at the Gulf of California. The Colorado is the most used river in the US. The basin is characterised by expansive and sparsely populated rural areas, with 80% of the population living in the rapidly developing urban centres of Tucson, Phoenix and Las Vegas.

The western part of the drainage basin is more arid than the east, with desert conditions in southern Utah and northern Arizona. Snowmelt from the mountainous headwaters provides nearly two-thirds of the annual runoff during late spring and early summer. Before the construction of dams the Colorado was a raging torrent in spring but little more than a trickle by the autumn. The Colorado is an unpredictable river due to natural climatic variability, which means that water resource managers must plan for long periods (years) with few floods and below average water supplies interspersed with short periods of high runoff.

Management
The river has a long and complex management history that dates back over almost 1,000 years to when prehistoric Indians first excavated canals in the valley of the Salt River in central Arizona. The first European colonists started to arrive in the mid-nineteenth century but development was slow due to the harsh environment. The US Bureau of Reclamation was established in 1903 to develop land through projects set up to

Resource 7: Map of Colorado Basin

Resource 8: Hoover Dam, Arizona – Nevada border

provide water for irrigation. In 1922 representatives from the seven Colorado basin states met to form the Colorado Compact which divides the basin into Upper and Lower parts (**Resource 7**) and assigned an equal amount of water to each. Following this, over 20 large multi-purpose dams, including the famous Hoover Dam (**Resource 8**), were constructed on the Colorado and its tributaries to regulate flow. The allotment of water to Mexico was not negotiated until a 1944 treaty. Today two million people in Mexico depend on the Colorado for their water supply. There have been numerous agreements since the 1922 Compact and all are collectively known as 'The Law of the River'. More water is potentially allocated on paper than the river can deliver in most years. The main reason for this is that the Compact was based on calculations of average flow for an unusually wet period and therefore over-estimated the potential supply.

The Colorado supplies water for irrigated agriculture, industry and urban centres within the drainage basin, which is supplemented by groundwater. In addition, water is exported from the Colorado basin to Denver, Salt Lake City, Southern California and to the upper Rio Grande basin in Mexico. The multi-purpose dams regulate these flows in addition to generating electricity, providing flood protection, fisheries and recreational facilities, and attracting tourism. Regulation also takes place verbally, according to 'The Law of the River'. There is increasingly severe competition for the small quantities of water that remain in the river and a considerable amount of litigation and controversy. Competing uses raise the cost of water, which is traded by water brokers as a commodity. Within the Colorado basin the question of water rights is as important as land ownership, if not more so, and owning land does not confer automatic entitlement to the water rights!

Water use and sustainability

The construction of dams and associated works has enabled the development of the basin, much of which has an arid or semi-arid climate. Between 1920 and 1960 the population within the basin quadrupled in size and today the cities of Tucson and Phoenix in southern Arizona are two of the fastest growing in the US. These cities were almost entirely reliant on groundwater until the mid-1980s when the Central Arizona Project (CAP) was completed. The CAP pumps water up from the main river over 1000 m high mountains and then 550 km of canals supply water-scarce regions in south and central Arizona. Phoenix and Tucson have the appearance of oases in the desert, with expanses of artificial lakes and parkland, numerous private swimming pools, green lawns surrounding houses and acres of golf courses. Vast quantities of water are used per capita by these cities, much of which is lost by evaporation. However, compared to agriculture, urban areas in the Colorado Basin use relatively little water.

The first dams on the Colorado provided a reliable source of water for irrigation and allowed the development of thousands of square kilometres of farmland. The main crops are citrus fruits, cotton and alfalfa, all of which are thirsty crops, but with government subsidies and a cheap and plentiful supply of water for irrigation there is little incentive to install more water-efficient irrigation systems. The enormous quantities of water that are used to flood fields, combined with very high rates of evaporation, have led to the age-old problems associated with irrigated agriculture in deserts: waterlogging and salinisation. As a result many thousands of square kilometers of farmland have been abandoned.

The current management structure provides little incentive to conserve water even though technologies exist. However, trading water as a commodity cannot create more of it and water availability will eventually become a limiting factor to further development. Even today the Colorado dries up before reaching its delta in the Gulf of California.

Chapter 4

Dynamic coasts and the challenge to human management

Why 'dynamic' coasts?

Differing time scales of coastal activity

The coastal zone is a dynamic geomorphological environment, which can alter every few seconds as waves hit the shoreline. However, people who live on the coast talk about changes stretching over their lifetime: and major change may take decades or centuries to accomplish. Coastal environments are thus characterised by short-term repetitive changes superimposed on long-term gradual changes. These short-term changes are an inherent feature of coastal processes, while longer term shifts are associated with changing external factors such as sea-level change; wave climate change; alterations to sediment supply; and even impacts of human activity that accumulate over generations. **Resource 9** shows this sequence of change with a gravel-dominated beach showing seasonal fluctuations in its extent. The gravel is often overlain by onshore sand transport

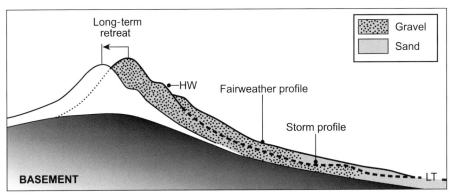

Resource 9: Seasonal (short-term) gravel beach profile change in comparison with longer-term (decade to century) beach ridge movement that is dependent on extreme storm overwashing

in fair-weather, while it is exposed by the action of storms that comb down the beach and remove sand off the beach face. If storms are more extreme, breaking waves may reach the beach ridge crest and push material on to the back slope. This slow, unsteady transfer gradually moves the gravel ridge onshore in a process called 'rollover', which can be a problem to landowners due to an irregular spread of gravel onto their land. They may rework the material back onto the ridge to slow down the process. However, with sea-level rise, there is an inevitable process of repetition of these overwashing and reworking events until the gravel ridge is so over-steepened (too high relative to its width) that when it is finally broken down by a major storm, the resultant scale of beach change is seen as catastrophic by landowners who fail to recognise their role in the process.

Coastal evolution and sea-level change

To understand coasts, one needs to consider their evolution since the last major glaciation when sea level was substantially lower. Over the last 20,000 years the Northern Ireland coast has seen relative shifts in land and sea as a consequence of isostatic (land rising after being depressed by ice) changes to land position and eustatic (climatic induced increases in the ocean's volume) changes to sea level. This fluctuating balance has caused a relative change in mean sea level over a vertical range of about 15 m. These up and down fluctuations occurred over thousands of years with the last major fluctuation showing a mean declining sea level rate of about 0.5 mm/year over the last 4000 years along eastern Ulster. The shoreline extending further to the sea preserved older beach ridges (Cushendun and Carnlough, Co Antrim) and where fine sand was available, formed the area for dune formation (Murlough, Co Down). Late twentieth century sea-level rise, thought to be associated with human-induced climate change, is a sudden reversal for Northern Ireland, the impact of which is as yet unknown.

Beaches assume the position, profile and alongshore shape that best absorbs the wave and tidal energy expended on them. These energy conditions are held to a physical elevation or datum set by mean sea-level at the shoreline. If mean sea-level rises as a result of climate change, then the whole beach system shifts landwards as wave and tidal energy reach further onshore in association with the rising sea level. This landward movement is hard for humans to resist without coastal defences, which in turn, if put in place, risk destabilising the onshore-moving beach system.

Sediment as a key to coastal change

During the last thousand years many beaches with low sea-level rise have also been associated with diminishing longshore sediment supply. As sediment supply reduces, existing beach material is reworked (cannibalised) and beach width reduced. Sand and gravel beaches are reliant for further sediment on longshore sources that have reduced in output as sea level stabilises. Human attempts to stave off cliff erosion have also reduced this supply. Fluctuations in mean sea-level can account for extra sediment being brought on to the beach and hence available for longshore transport, especially under storm conditions. Fluctuation in sea level over time can lead to alternating periods of beach sediment scarcity and excess. The cooler and wetter conditions of the fourteenth to eighteenth centuries (known as the Little Ice Age) may have been associated with the last major period of extra beach sediment at times of lower than present sea level. Sediment scarcity is a major factor behind many coastal erosion problems of modern times.

Sediment movement onshore is the likely source for protected estuaries where sub-tidal banks and high tidal marshes are accreting through offshore-sourced mud supplies. There is a strong relationship between sediment size and available marine energy. Gravel and sand beaches are dominated by waves while the fine sediments of upper estuaries are controlled by tidal action.

Sea level rise will push coastal environments onshore as long as they are not interrupted by human activities. **Resources 10A & B** show before and after views of a low-lying coastal barrier island in North Carolina, USA, that is being pushed landwards by a 1–2 mm/yr sea-level rise. The natural barrier (**Resource 10A**) shows a wide beach and low

Resource 10: Comparison of a natural barrier (A) with a stabilised barrier (B), North Carolina, USA. The main difference is the reduced beach width on the stabilised barrier. This allows bigger waves closer inshore before breaking and will eventually overwhelm the stabilisation unless larger defences are built

back-barrier with overwashed storm sediment inter-fingering with marsh to form a natural foundation for future beaches that move onshore with rising sea level. **Resource 10B** shows a nearby barrier that has been fixed temporarily by building an artificial dune line to protect the road from overwash. Note the reduced beach width as beach sediment has been moved longshore rather than going onshore. The barrier is narrowing, but at some future date the dunes will not be sufficient to prevent overwash as bigger waves get closer to the beach as it declines in width. At such a point 'someone' will have to decide whether the infrastructure is more important than the barrier and continue with more drastic measures to protect the road.

Beaches are vulnerable to alteration as sediment supply diminishes. The key issue to resolving coastal erosion is in supplying more sediment to maintain an adequate beach buffer (high beaches) to wave conditions. Lower beaches mean bigger waves that reach further onto the land, causing increased erosion and greater damage to property.

Human perceptions of coastal erosion rates

Rates of erosion averaging <0.5 m/yr are not uncommon around the British Isles; rates of >1 m/yr are thought of as being problematic by local government. The problem is that average rates can disguise the fact that a single severe storm may cause several years' worth of average damage. However, as map data of chalk cliff retreat rates at Birling Gap, near Beachy Head and Seven Sisters, southern England (**Resource 11A & B**) show, such dramatic episodic storm events do not always mirror the actual long-term coastal trend. People can overestimate how dynamic a coast might be by taking a short term view. However, the key message for people living by the coast is that they should expect some degree of coastal instability.

Hard and soft coastal protection

The engineering trap

People living in perilous conditions at the shoreline expect coastal protection to be built that reinforces safety and property survival. Engineers are more than willing to build

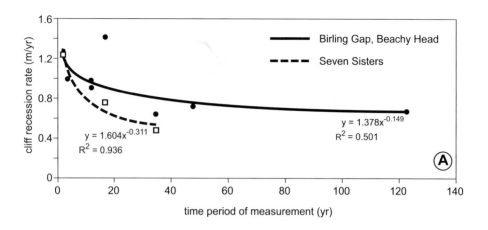

Resource 11: Recession rates (A) measured on chalk cliffs (B) as a function of observational time base, Birling Gap, Sussex, near Beachy Head. There is a negative exponential relationship that indicates assessment over 30 years or less overemphasises the century rate of retreat, thus human memory does not always reflect nature's way. This underlines the importance of checking people's perspective of coastal threats.

defences (hard protection) to achieve this end, but may not accept that their efforts (regardless of technological ability) rarely achieve complete success. At best the rates of coastal retreat are reduced while at worst their efforts make no difference in the long run. Each generation appears to have to learn salient truths: that committing yourself to

Resource 12: Hard coastal protection: (A) Victorian sea wall plus recent additions at Morecambe, Lancashire; (B) contemporary equivalent at Selsey Bill, Sussex; (C) groyne field at Eastbourne, Sussex

engineered coastal defence means you are never free from engineering; that the cost of engineering becomes more than the value of what you protect; and that engineering may protect the landward side but destroys the seaward side of the coast.

Sea walls: hard protection

Despite the long Dutch history of coastal engineering, modern approaches start with the engineers building protection for burgeoning Victorian coastal resorts and harbours. Victorian engineers saw it as a civic duty to design structures to protect society from the forces of nature. Structures were solid and immovable, hence the generation of massive stone-built sea walls at coastal resorts like Blackpool and Morecambe (**Resource 12A**) that supported the promenades for visitors to 'take the air'. Such sea walls were designed to protect property from the direct force of the breaking waves. Walls were an obvious extension of the Dutch dyke or clay banks that had been used in estuaries and marshes for many centuries, solely to prevent marine flooding of agricultural land. These dykes relied on the seaward intertidal banks and salt marshes to dissipate the breaking wave's energy before it reached the clay bank as such.

Unfortunately, sea walls carried their own seeds of self-destruction. Reflected waves off the wall interacted with incoming waves and created a turbulent water motion leading to a scouring of sediment from the beach. As beach volume fell, a sea wall's foundations were exposed and undermined. **Resource 12A** shows that the beaches at Morecambe (Lancashire) have now disappeared due to reflection, and large boulders have been laid down at vulnerable points in front of the wall to dissipate wave energy. The lack of beach means that larger waves break closer to the shore and in extreme storms overwash the promenade. A floodwall built at the rear of the promenade now protects the basements of hotels. The high cost of building seawalls (c£1000/m) and the associated loss of beach means that walls are not cost/benefit efficient. This does not stop many private individuals from attempting to build them or pressuring local authorities to support such ventures (**Resource12B** shows individualistic attempts to live with a sea wall at Selsey Bill, Sussex.)

Groynes: hard protection

The maintenance of beach sediment is essential in coastal protection. Whereas walls protect property directly, groynes are designed to maintain beach sediment. Groynes are concrete (formerly wooden) fences that are built across to the shore from above the high tide position out onto the intertidal area and are designed to stop the movement of longshore sediment and form a buffer to wave energy. **Resource12C** shows a recent groyne field (multiple groynes) built at Eastbourne, southern England. Their obvious presence is some indication of a lack of success, in that successful groynes should not be seen but rather be buried under a beach volume that still allows some surface transport (ie above the buried groyne) to feed down drift beaches. Groynes present a paradox, in that successful groynes require a longshore supply that, if sufficient, should not need groynes in the first place.

Groynes create problems in that, until buried, they stop any protective sediment from reaching further down drift positions and generate 'terminal scour' or beach erosion after the last groyne. The adage 'once you start you cannot stop' is particularly true for groynes. **Resource 13A** shows Sandy Hook, New Jersey, where groynes were first built to stem beach loss (and protect urban development), then a sea wall, which is now rock armoured. All that is left 60 years on are pathetic fragments of the beach that the houses were built to look over. This divorcing the community from the shore by extreme protection is known as 'Newjerseyisation'. **Resource 13B** shows an early stage in this process at Long Beach Island, New Jersey, with numerous groynes built to curtail longshore sediment loss. With no room for urban retreat, this island is a candidate for containment by sea walls given future sea-level rise, assuming acceptance of the high costs involved.

Resource 13: Effects of hard protection beach depletion: (A) Sandy Hook, New Jersey; (B) Long Island Beach, New Jersey

Resource 14: Beach nourishment at Miami Beach, Florida: (A) beach depletion in 1972, (B) after nourishment in 1980

Beach nourishment: soft protection

Built protection costs increase geometrically as protection height grows linearly. The spiralling costs of and environmental damage from hard protection methods has forced a search for alternative soft protection methods. The USA has led the way in using a less costly and less environmentally damaging approach: if a beach is missing, replace the beach rather than build protective structures. **Resource 14** shows the before (A: 1972) and the after (B: 1980) views of Miami Beach, where the financial scale ($1 billion) of apartments and hotels exposed to erosion outweighed the $10m cost of nourishing 16 km of beach (1980). The success of this project means that nourishment is now undertaken on a massive scale as the only logical soft alternative to hard and intrusive built engineering structures. However, there is growing concern that beach nourishment is not a long-term solution (nearly 50% of US projects failed within 10 years – often due to extreme storms) and is being used unwisely. If a beach has disappeared, replacing it without asking why the beach disappeared in the first place is not good practice.

Case study: coastal protection in West Sussex

West Sussex (**Resource 15**) is a low-lying area of periglacial and interglacial silt to gravel sediment banked against the rising chalk of the South Downs to the north and east. As sea level has risen over the last 8000 years the coastline has moved landward over this easily eroded sediment, exposing the underlying chalk platform to the east at Beachy Head (**Resource 11B**). The lack of fine sand in the eroding landward sediment, as well as in the offshore zone, is reflected in the concentration of gravel (shingle) in the beaches. The prevailing eastward moving waves have moved much of the beach sediment to the east; indeed some believe that the giant beach ridge plain (cuspate foreland) at Dungeness is primarily composed of beach material derived from this coastline. The power of this eastward longshore drift is seen in the growth of spits at Shoreham and Seaford, which have been cut through and stabilised to allow access to port facilities. The consistent easterly drift of beach sediment means that most beaches are deficient in sediment. It is a natural response that gravel beaches with low beach volume tend to form a beach ridge at the high tide position. This ridge is the

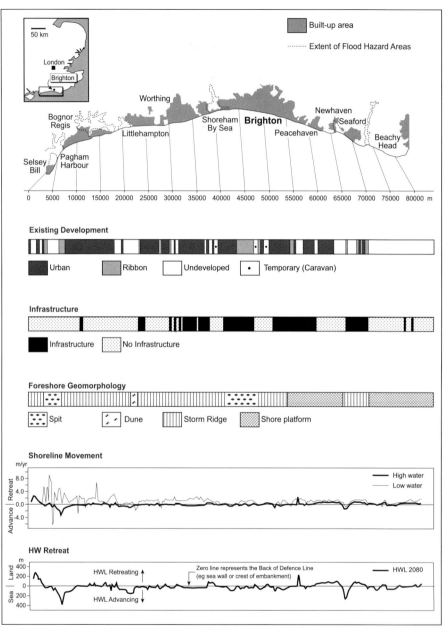

Resource 15: Coastal zone structure of West Sussex

only natural bulwark to flooding from storm wave activity, and retreats landwards via rollover, as sea level rises.

The coastline was the site of several small fishing villages, which became popular with Georgian and Victorian seaside tourists. It was George IV's fascination with one in the late eighteenth century that effectively started the trend of coastal resorts: that village is now the urban area of Brighton and dominates the eastern end of this shoreline. Georgian development was generally set back from the coast as they were aware of coastal flooding potential. It is unfortunate that later Victorian developers were not so aware and rushed into developing these safety zones; shoreline resort development was a significant element of the county's economy, though in a hazardous way. Sussex is the classic case of rapid Victorian coastal population expansion post-1860, much of it linked to the opening of the railway between Chichester and Hastings. Twentieth-century suburbanisation based on the image of the coast as a source of pleasure, quickly infilled gaps between the older resorts. Places like Peacehaven sprang up as unplanned residential communities that grew in a ribbon-like process along the shore. Each new community added to the growing pressure to maintain a fixed shoreline. Although sea walls as promenades were evident in the major towns, eg Brighton, groynes have been the dominant hard engineering shoreline defence used to stem the easterly-directed longshore drift. In most cases once groynes were started, it effectively stopped further sediment supply to the east, thus necessitating even more groynes. As beach sediment volumes fell, storm ridges were overtopped. It was a common practice to push rollover material back to the ridge crest, thus over-steepening the ridge and ensuring further difficulties as the beaches could not retreat when sea level rose. By the late twentieth century, the lack of beach material was so obvious that beach gravel renourishment became a constant need between Bognor and Brighton. Now after every major storm season, rebuilding the seaward beach profiles has to be undertaken (soft engineering). The beaches are renourished about every five years and then the material is remoulded by bulldozers as necessary until more is needed. Vulnerable points, such as Selsey Bill, might need renourishment every year. Soft engineering costs about 25% as much to install as hard engineering, sometimes less. In 1982 sand renourishment at Bournemouth cost £350 per metre, when a piled wall would have cost £1900 per metre. The West Sussex coastline has become synonymous with the maintenance of hard and soft protection that is a direct charge to central government resources (all society pays) but which brings a benefit to only a few coastal land and property owners. Is this an equitable exchange? It certainly is not a sustainable one as few of these beaches now show a natural structure that can be maintained for future generations. There used to be, too, the environmental costs of extracting gravel for beach protection, which came largely from Sussex river valleys, but now most has to be dredged from offshore banks. West Sussex has accepted the present coastline as one dominated by hard and soft coastal engineering, without which the coastline's continuity and coherence would rapidly disintegrate.

Management strategies for the future
Coastal protection as a cost/benefit problem for society
Coastal protection is generally costly and damaging to the coastal environment. It may have short-term benefits for those few at the shoreline but the cost (estimated £75 m to £225 m per annum for the UK) is a burden on all society and for little benefit. Coastal property values may rise, but there is no major return for the rest of society, which also carries the cost of environmental destruction. What then can society do in this situation?

Options for coastal management
There are four generally recognised management strategies to deal with coastal protection issues.

- *Hold the line.* Maintain engineered protection for all coastal problems. A costly solution and one only likely where political will and fiscal ability are sufficient to support this option in the face of a threat to the very existence of life and society, eg The Netherlands. Island states in the South Pacific, such as Tuvalu, whose existence are equally threatened by rising sea-level, do not have the resources to pursue this option and may have to consider the relocation of their population.

- *Do nothing.* There may come a point when coastal problems and costs reach a scale and complexity that a 'do nothing' attitude prevails whether by default or design. The massive scale and severity of coastal problems in Bangladesh suggests that the 'do nothing' option has been operating there by default for some time.

- *Accommodation.* This is where society recognises that some coastal elements will be defended while others will not. To some extent the distinction will be based on legal requirements and pragmatic coastal scale considerations. Along eastern England are numerous coastal habitat reserves that have to be protected (by European habitat law) or, if eroded, new sites have to be provided. It is likely that coastal infrastructure and urban areas will be considered for continuing protection, though speculative retirement estates and individual coastal houses are already proving to be controversial with respect to UK protection requests. It is unfortunate that the number of sites requiring future coastal protection is growing all the time in the UK, due to a planning system that is unable by law to deny individual coastal development, regardless of future hazardous problems it may face. The extreme example of this is UK coastal nuclear power stations whose working life is nearly up, but whose level of radioactivity requires the reactor cores to be defended from the sea for several hundred years, until radioactivity diminishes sufficiently for safe dismantling.

- *Retreat (shoreline realignment).* Retreat is a highly emotive issue to societies unprepared it, yet accelerating sea-level rise and spiralling protection costs identify long-term retreat as the only viable alternative when dealing with shorelines of low value. The expression 'shoreline realignment' is a more neutral label for shoreline retreat. The few cases where retreat has been undertaken reflect an uncertain science in predicting future shorelines. The south and east UK coasts are now so heavily defended that a policy of retreat is made more uncertain by lack of clarity as to what the natural shoreline is, let alone predicting any future one.

Resource 15 also illustrates the problematic future of the urbanised west Sussex coast.

Coastal walls, groynes and, latterly, constant gravel recharging of beaches to maintain the shingle storm ridges have held the shoreline. The high-water line has remained static over the last century despite the low-tide line moving shoreward. This steepening of beach slope could be fatal given future increased English Channel storminess. Inevitably future periods of beach overtopping and shoreline retreat due to rising sea level will be accompanied by demands to maintain and raise protection levels. The issue is whether such demands can or should be fulfilled, and who pays – shoreline property owners or society through general taxation.

Clearly the built investment is too great to be left to sea attack and for the foreseeable future this coast's defences will be maintained – doing nothing is not likely, so holding the line will prevail. Logic suggests that there will be continuing stabilisation of existing beach ridges to the point that they will be purely artificial and heavily dependent on constant sediment recharging or else protected by major rock-armouring, which will further steepen the beach slope and open the way for bigger waves to attack. Realignment is improbable at the moment, given the lack of space for the shoreline to retreat. It is unfortunate that any temporary gain from holding the line will not be associated with any control on the built development by which a gradual downward shift in economic values of coastal property could be pursued, thus paving the way for a logical future coastal retreat. Unfortunately coastal defences tend to enhance rather than depress property prices.

The lack of overall coastal management means that a local, narrow, partisan, perspective of 'saving my property at someone else's expense' will be maintained, until such time that the threats and responses make managed realignment the only viable option. **Resource 16** shows the UK government's view that within 20 years realignment is likely to be the preferred pathway to effective coastal management.

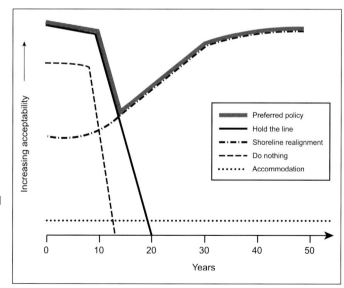

Resource 16: Relative likelihood of coastal management options being implemented over the next 50 years

Sand dunes as a sensitive environment to human interaction (incorporating a case study of the dunes at Murlough, Co Down)

Structure and formation of dune ecosystem

Formation of coastal dunes

Coastal dunes are formed from supply of fine-medium beach sand (<0.2 mm) that dries sufficiently to be transported by an onshore wind into a back-beach area, where the dunes can form free from storm-wave attack. Wind velocity has to increase at a geometric rate (cube power) to transport sediment sizes increasing linearly. Extreme storm wind is required to transport larger than medium sand. Coarse sand can be transported when suspended grains fall and strike surface grains that then move (saltation).

Coasts have organic and, in modern times, inorganic litter stranded at the storm-tide line — seaweed, shells, drift wood, plastic containers — sufficient to act as obstacles. Any obstacle to the wind flow will cause deposition of blown sand. The wind flow is compressed immediately above the litter and then expands, dropping velocity and suspended sediment down-wind. Sediment deposited around the litter forms an incipient dune that acts as an obstacle to further transport and causes a foredune to form. A series of foredunes is often associated with an influx of beach sediment that first progrades (builds the beach seawards) the beach and forms the foundation for new foredunes to build seawards. Foredunes can also accompany falling sea level that provides both fresh beach sediment and space for dunes to develop in. This combination is why Murlough dunes formed during the last 6000 years (prograded dunes).

Coastal dunes are different from desert dunes due to precipitation and vegetation. Free sand surfaces in coastal dunes can be stabilised by vegetation, whereas desert dunes cannot. Coastal dune surfaces tend over time to show a succession of plants, starting as salt tolerant and dependent on continuing sand deposition (eg marram grass: **Ammophila arenaria**). Over time, as new dunes form to seaward, the supply of sand to the landward dunes is reduced, rainfall leaches salts out of the surface layers, nutrients build-up from decaying plant remnants, and marram gives way to grasses. The dune surface starts to stabilise so that over time a well-drained sandy soil develops, which is capable of supporting a continuous sward of low grasses. In low points between dunes (slacks) the water table will be closer to the dune surface, even exposed in wet periods, and form habitats for shrubs and trees, assuming non-waterlogged conditions. This succession of dune plants is well established in most coastal dunes and explains why marram is only located at points where fresh sediment is most likely, eg blow-outs, or at the dune–beach boundary. Variation in rainfall, nutrient provision and surface stability can create different linkages in the dune ecosystem. Shell fragments (carbonates) are often included in dunes. Precipitation leaches carbonates out and, without replacement, dunes turn acidic and support strong communities of heath-like plants such as heather and bracken.

Change in coastal dunes

Constant onshore wind can still move sediment through the foredunes by means of cross-dune channels or blowouts and trigger off the growth of landward dunes. Blowouts may arise from animal track ways, natural dips, and pedestrian tracks. If the dune's vegetated surface is stressed — by trampling, reductions in precipitation or over grazing by wild or domesticated animals — then the root and litter binding is reduced and dune sediment exposed to reworking by the wind. Evidence of major coastal dune reworking is shown by the presence of parabolic dunes that have climbed over the older lower dunes to form younger, higher dunes (**Resource 17A**). Murlough has evidence of such higher, younger dunes occurring in the Little Ice Age, overlying older dune surfaces 2000–5000 years old.

Accelerated rising sea level will erode the seaward edge of the dunes. New dunes are only likely to occur down drift from the eroding site where accommodation space and local progradation is available. The dune ecosystem could be stressed by changes in rain, wind and ground water-table position accompanying the climate change responsible for sea-level rise.

Impact and remedial action associated with human pressures on dunes

Historically dunes have been viewed as worthless. In the twelfth century Normans introduced rabbits (for winter meat) into Ireland and created warrens or protected dunes for them. Periodically rabbit populations expanded beyond the vegetation carrying capacity of the dunes and overgrazed the surface vegetation, causing surface instability and reworking. Reworking is a common aspect of dunes though in recent times it has been viewed, without reason, as undesirable. Golf course development is seen as a way of stabilising dunes and generating value from an otherwise undervalued resource.

Twentieth century recreational pressures have been felt on dunes via trampling and associated blowout development. This caused sufficient concern for past management practice to promote reduction in access and stability of the dune surface to the point of fossilisation. The complete stabilisation of dunes is not now viewed as the sustainable way to manage dunes. Murlough dunes became stabilised after three decades as a National Nature Reserve (**Resource 17B**), whose policy emphasised prevention of public access. Recent efforts have been made to reintroduce rabbits and other domesticated animals to ensure grazing opens up areas to sand reworking, and remove the tendency for bracken in particular, to crowd out other dune grassland species. The reworking of sand is now seen as an essential element of a sand dune; however, the difficulty is gaining a balance between active and stabilised dunes. In the USA the argument for stabilisation is taken to such lengths that crossing dunes to access a beach is often by means of complex elevated boardwalks or walkways. In this

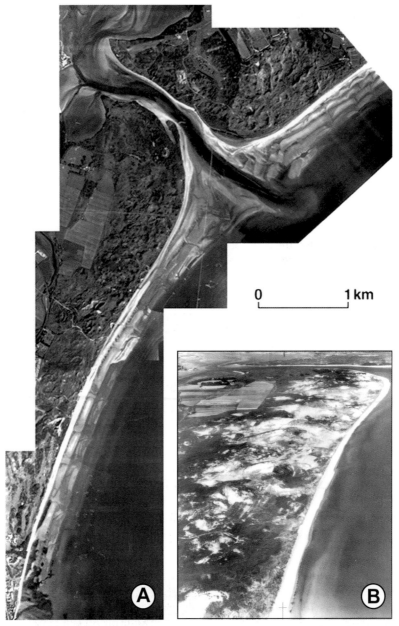

Resource 17: Murlough Dunes, Co Down: (A) Vegetation-free dunes in 1956.
Note the arc of several parabolic dunes directed to the north-east, from Little
Ice Age reworking; (B) Murlough vegetated and stabilised in 1999 after 30
years free from human access

way people are not allowed to walk on the dune surface for fear of disturbing its stability (**Resources 18A & B**).

Dunes are often seen as a good defence against storm waves as long as sediment can be exchanged between dune, beach and dune. Wave eroded sediment from dunes helps to defend beaches during storms, while beach sediment returns to the dunes at times of non-storm conditions. Home-owners living on US coastal barriers have great faith in the presence of dunes, to the point where they attempt to initiate dunes with sand fences and sow marram-type vegetation to stabilise any collecting blown sand regardless of the lack of accommodation space (**Resource 18C**).

Defending dunes from coastal erosion is not helpful to the coastal system's balance. Coastal golf courses (links courses) are keen to maintain their seaward boundaries by protection. Hard protection dune defence is not unusual and cheaper versions of sea walls (rock armouring) are becoming common (eg Murlough). However, it is costly, rarely beneficial and closes down the required beach/dune exchange for dune maintenance. The only effective form of dune protection is a plentiful beach sediment supply. In most cases around the Irish coast, the lack of current fresh dunes indicates the major shortage of beach sediment and the likelihood of continuing dune erosion with twenty-first century sea-level rise.

Resource 18:

(A) Cross-dune boardwalks at Miami to prevent access trampling of fragile rebuilt dunes

(B) Private house boardwalk access on fragile dunes, Fire Island, New York

(C) Sand fences to encourage dune development but in an inappropriate position, too close to the high-water line

Bibliography and further reading

Brookes, BA (1985) 'River channelisation: traditional engineering practices, physical effects and alternative practices' *Progress in Physical Geography*, pp 9, 44–73

Brookes, AB (1987) 'River channel adjustments downstream from channelisation works in England and Wales', *Earth Surface Processes and Landforms*, pp 12, 337–355

Clayton, K (1989) *Coastal geomorphology* Macmillan, London

French, P (2001) *Coastal Defences; Process, problems and solutions* Routledge, London

Gleick, PH (1993) *Water in crisis: a guide to the world's freshwater resources* Oxford University Press, Oxford

Gordon, ND, McMahon, TA and Finlayson, B (1992) *Stream hydrology: an introduction for ecologists* Wiley, London

Graf, WL (1985) *The Colorado River: instability and basin management* Association of American Geographers, Washington, DC

Haslett, S (2000) *Coastal Systems* Routledge, London

Keller, EA (1976) 'Channelisation: environmental, geomorphological and engineering aspects', in DR Coates (Ed) *Geomorphology and engineering* George, Allen and Unwin, London

McDonald, AT and Kay, D (1988) *Water resources: issues and strategies* Longman, London

Jones, JAA 1997) *Global hydrology* Longman, London

Manning, JC (1997) *Applied principles of hydrology* Prentice Hall, Harlow

Newson, MD (1992) *Land, water and development: river basin systems and their sustainable management* Routledge, London

Newson, MD (1994) *Hydrology and the river environment* Oxford University Press, Oxford

Park, C (2001) *The environment: principles and applications* 2nd edition, Routledge, London

Petts, G and Calow, P (Eds) (1996) *River restoration* Blackwell, Oxford

Thorne, CR, Hey, RD and Newson, M (1997) *Applied fluvial geomorphology for river engineering and management* Wiley, London

Vivash, R (1999) *Manual of river restoration techniques* The River Restoration Centre, Arca Press, Bedford

Wilcock, DN and Essery, CI (1991) 'Environmental impacts of channelisation on the River Main, County Antrim, Northern Ireland', *Journal of Environmental Management*, pp 32, 127–43

Useful websites

http://www.therrc.co.uk/ River Restoration Centre. General information on river restoration, data on specific projects and past newsletters.

http://www.waterinfo.org/ US Southwestern Water Conservation District. Information on water conservation, regional water projects, Colorado water; Law of the River, history and links to other sites.

http://www.uc.usbr.gov/ and http://www.lc.usbr.gov/ US Bureau of Reclamation Upper and Lower Colorado Regional Offices. Information on management, operations, data, environmental programmes and research.

http://ag.arizona.edu/AZWATER/main.html College of Agriculture and Life Sciences, University of Arizona. Summary of water resources issues in Arizona; Colorado River, Water Rights, Conservation, Climate, The Central Arizona Project, Water Uses.

http://www.environment-agency.gov.uk/ UK Environment Agency. General information and maps on both land and coastal flooding for England and Wales.

Video

River Restoration Centre (1998) *Rivers of the future: restoring the Cole and Skerne, UK, 1995–97*. RRC, Cranfield. 30 mins. Available from The River Restoration Centre, Silsoe Campus, Silsoe, Beds. MK45 4DT, E-mail: rrc@cranfield.ac.uk (web address above).

Acknowledgement:

Rosemary Charlton would like to thank Michael Oliver, Conservation Officer at the Rivers Agency, Belfast for providing maps, photographs and information on the Ballysally Blagh Project; also thanks to James Keenan of the Geography Department in NUI, Maynooth.